FANTASTIC PLANTS

STINKY CORPSE FLOWERS

MARY GRIFFIN

PowerKiDS press.
New York

Published in 2023 by The Rosen Publishing Group, Inc.
29 East 21st Street, New York, NY 10010

First Edition

Portions of this work were originally authored by Tayler Cole and published as *Corpse Flowers Smell Nasty!* All new material in this edition was authored by Mary Griffin.

Editor: Therese Shea
Book Design: Michael Flynn

Photo Credits: Cover Aleksandra H. Kossowska/Shutterstock.com; back cover, interior background (illustration) milhad; interior frame Liubou Yasiukovich/Shutterstock.com; p. 5 (corpse flower) Isabelle OHara/Shutterstock.com; p. 5 (map) Porcupen/Shutterstock.com; p. 7 Erik Cox Photography/Shutterstock.com; p. 8 https://commons.wikimedia.org/wiki/File:Amorphophallus_Corm_0136a.jpg; p. 9 https://commons.wikimedia.org/wiki/File:Amorphophallus_titanum,_young_leaf,_Bonn._Foto_%C2%A9_W._Barthlott,_Bot. Gard._Bonn.jpg; p. 11 Denise Kappa/Shutterstock.com; p. 13 (korm) Morphart Creation/Shutterstock.com; p. 13 (plant) Bodor Tivadar/Shutterstock.com; p. 15 Phillip B. Espinasse/Shutterstock.com; p. 16 Jamil Bin Mat Isa/Shutterstock.com; p. 17 syahrilinggilu/Shutterstock.com; p. 19 Moritz Fraisl/Shutterstock.com; p. 21 Erifqi Ze/Shutterstock.com.

Library of Congress Cataloging-in-Publication Data

Names: Griffin, Mary, 1978- author.
Title: Stinky corpse flowers / Mary Griffin.
Description: New York : PowerKids Press, [2023] | Series: Fantastic plants | Includes index.
Identifiers: LCCN 2021052212 (print) | LCCN 2021052213 (ebook) | ISBN 9781538386637 (library binding) | ISBN 9781538386613 (paperback) | ISBN 9781538386620 (set) | ISBN 9781538386644 (ebook)
Subjects: LCSH: Amorphophallus–Juvenile literature.
Classification: LCC QK495.A685 G75 2023 (print) | LCC QK495.A685 (ebook) | DDC 584/.442–dc23/eng/20211101
LC record available at https://lccn.loc.gov/2021052212
LC ebook record available at https://lccn.loc.gov/2021052213

Manufactured in the United States of America

CPSIA Compliance Information: Batch #CSPK23. For Further Information contact Rosen Publishing, New York, New York at 1-800-237-9932.

CONTENTS

WHAT'S THAT SMELL?

Indonesia is a beautiful country in Southeast Asia. If you're exploring a rain forest on the Indonesian island of Sumatra, you might come across an awful, rotting smell. Don't worry—it's not something dead. It's the scent of a flower!

When it blooms, the corpse flower smells like rotting flesh. Scientists believe the scent draws **insects** such as beetles, bees, and flies to the plant. The bloom lasts less than two days. You're lucky if you get a sniff of the gross smell!

BE AN EXPERT!

THE NAME "CORPSE FLOWER" COMES FROM THE INDONESIAN WORDS *BUNGA* ("FLOWER") AND *BANGKAI* ("**CARRION**").

THE CORPSE FLOWER IS KNOWN FOR ITS SIZE. IT'S ABLE TO GROW MORE THAN 10 FEET (3 M) TALL!

HUMID HABITAT

The rain forests of western Sumatra, Indonesia, are located close to the **equator**. That means they're very warm all year round. Rain forests get a lot of rain too. This is a perfect **habitat** for corpse flowers. They like moist soil, or soil that's a bit wet.

The corpse flower is often found in clearings on hills in the rain forest. It needs a lot of sunlight. The large plant also needs a lot of space around it to grow and bloom.

BE AN EXPERT!

CORPSE FLOWERS HAVE ANOTHER NAME: TITAN ARUM. "TITAN" IS A WORD THAT MEANS "ONE THAT IS VERY LARGE IN SIZE."

MANY CORPSE FLOWERS ARE FOUND IN INDOOR GARDENS CALLED BOTANICAL GARDENS. THEY'RE GROWN IN ROOMS THAT MIMIC, OR COPY, THE HEAT AND **HUMIDITY** OF A RAIN FOREST HABITAT.

FROM SEED

The corpse flower starts as a seed. It grows a **tuber** underground called a corm. The plant produces one large leaf at a time. The leaf, which lives for up to two years, makes food for the tuber through **photosynthesis**. Every time the plant grows a new leaf, the corm gets bigger. After 10 years, the tuber may be larger than a basketball.

Most years, the plant only grows a leaf. Its first flower can take 7 to 10 years to appear.

BE AN EXPERT!

SOME CORPSE FLOWERS ONLY BLOOM ONCE EVERY 10 YEARS. OTHERS BLOOM MORE OFTEN, BUT STILL NOT EVERY YEAR.

THE LEAF OF THE CORPSE FLOWER LOOKS LIKE A SMALL TREE. IT CAN GROW MORE THAN 20 FEET (6 M) HIGH.

TIME TO FLOWER

The corpse flower gets ready to bloom over several months. It has a tall part in the middle called the spadix. A large, petallike part called the spathe surrounds the spadix. Inside the spathe, at the bottom of the spadix, are tiny flowers.

When it's nearing the time to bloom, the plant can grow as much as 6 inches (15 cm) a day! Its temperature rises, and the flower begins to give off its terrible smell. The corpse flower is said to bloom when the spathe opens.

BE AN EXPERT!

THE SPATHE IS NOT THE FLOWER. THE PLANT'S ACTUAL FLOWERS CAN'T BE SEEN EASILY WHEN IT'S IN BLOOM.

THE INSIDE OF THE SPATHE IS A MAROON COLOR.

IN BLOOM

The corpse flower commonly stays open for about 24 hours. It can begin blooming in the afternoon. The smell increases throughout the night. Then the bloom and smell may be gone by the next afternoon. Flowers grown in greenhouses often bloom a bit longer.

The corpse flower gives off a smell made up of over 30 **chemicals**. The stink is for a reason—to attract, or draw, bugs to the plant. Its tiny flowers need to be pollinated for the plant to **reproduce**.

BE AN EXPERT!

SOME CHEMICALS IN THE CORPSE FLOWER'S SCENT ARE THE SAME AS THE CHEMICALS GIVEN OFF BY SMELLY CHEESE, ROTTING FISH, AND SWEATY SOCKS!

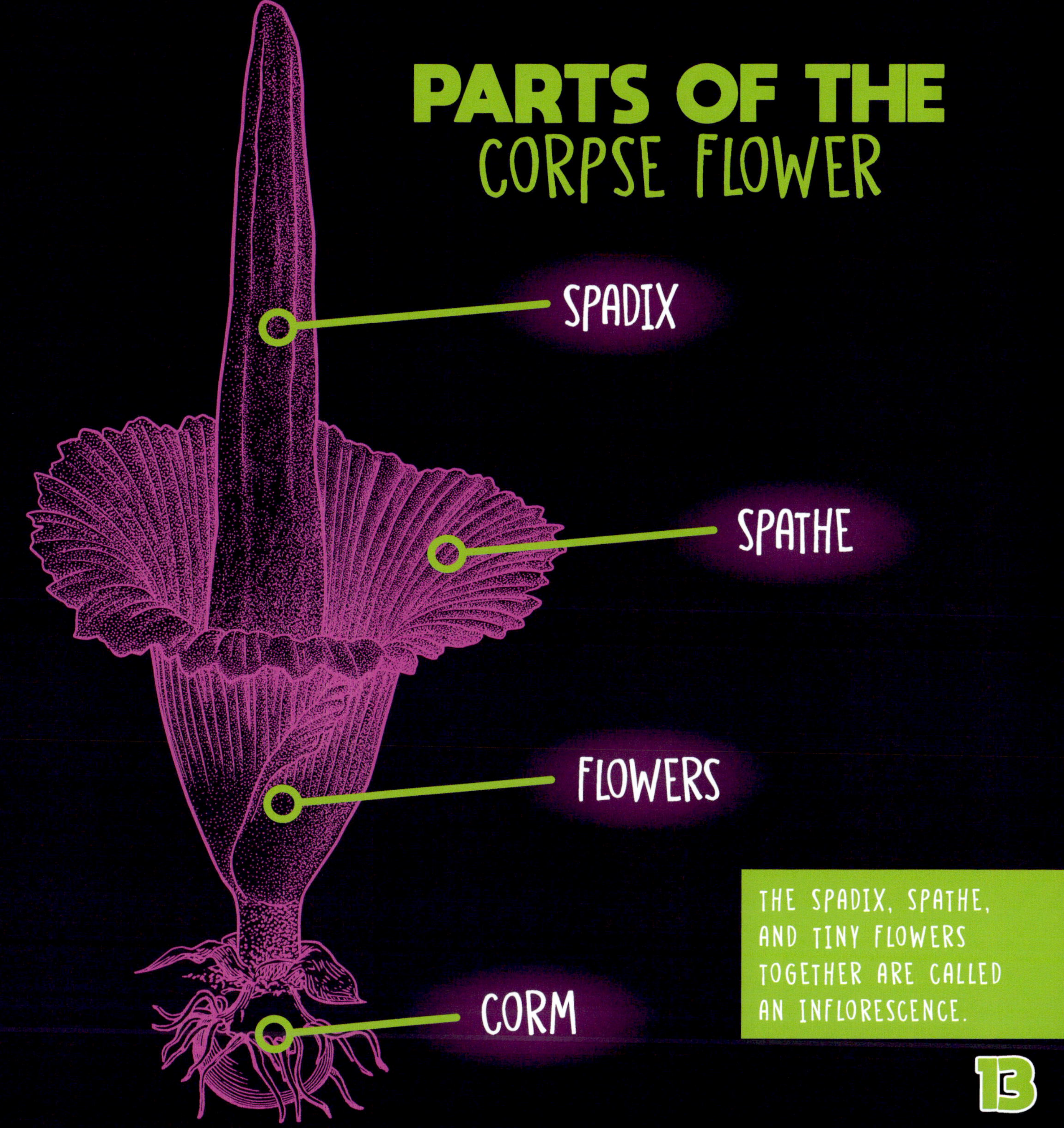
PARTS OF THE
CORPSE FLOWER
SPADIX
SPATHE
FLOWERS
CORM
THE SPADIX, SPATHE, AND TINY FLOWERS TOGETHER ARE CALLED AN INFLORESCENCE.

WELCOME, POLLINATORS!

The hundreds of tiny flowers at the base of the spadix are split into two groups. Half are female flowers, and half are male flowers.

When the spathe opens, the female flowers are sticky and ready for pollen from another plant. The male flowers won't **release** their pollen until hours later, after the plant's female flowers can no longer be pollinated. This is a special **adaptation** of the plant. If the plant pollinated itself, it wouldn't produce strong new plants.

BE AN EXPERT!

POLLINATION IS PART OF **FERTILIZATION**. DURING THE PROCESS OF POLLINATION, POLLEN IS MOVED FROM ONE FLOWER TO ANOTHER, OFTEN ON THE LEGS AND BODIES OF BUGS.

IN THIS PHOTO, YOU CAN SEE THE YELLOW MALE FLOWERS AND ORANGE AND PURPLE FEMALE FLOWERS BENEATH THEM.

BERRIES AND SEEDS

After the female flowers of the corpse flower are pollinated, the plant grows a tall stalk of berries over several months. First, the berries are yellow. When they're ripe, they turn red. The berries have seeds inside. They're eaten by animals, such as the bird called the rhinoceros hornbill. The seeds are spread throughout the rain forest in the animals' waste.

After this, the plant becomes dormant for a year or so. Then, it will grow leaves again until it stores up enough energy, or power, to bloom.

RHINOCEROS HORNBILL

BE AN EXPERT!

A PLANT THAT IS DORMANT IS NOT ACTIVELY GROWING, BUT IT'S STILL ALIVE.

AFTER THE FRUIT OF THE CORPSE FLOWER IS FULLY GROWN, THE SPATHE FALLS OFF.

DISCOVERY

Italian botanist Odoardo Beccari was the first scientist to write about the corpse flower. He observed it during a trip to Indonesia in 1878. He sent seeds and corms of the plant back to Europe. The corms didn't survive, but one seed did grow into a plant. Its first bloom was observed in London, England, in 1889.

Since then, the corpse flower has been grown in botanical gardens throughout the world. Every year visitors somewhere in the world line up to see—and smell—a corpse flower blooming.

BE AN EXPERT!

A BOTANIST IS A SCIENTIST WHO STUDIES PLANTS.

SOMETIMES THOUSANDS OF PEOPLE VISIT A SINGLE CORPSE FLOWER TO EXPERIENCE THE STRANGE SIGHT AND SMELL!

DISAPPEARING

Because the corpse flower's bloom doesn't happen often, each bloom is an exciting event. Imagine taking care of a plant for 10 years, just to see it bloom once!

However, this plant is in danger of dying out in its native habitat. Much of its rain forest home in Sumatra has been cut down to clear land for plantations, or large farms. Clearing the rain forest also harms the animals needed to spread the corpse flower's seeds. Scientists are working hard to keep this amazing plant blooming.

BE AN EXPERT!

THE RECORD FOR LARGEST SINGLE FLOWER BELONGS TO ANOTHER PLANT IN INDONESIA, THE MONSTER FLOWER (*RAFFLESIA ARNOLDII*).

THE CORPSE FLOWER'S CORM IS A FOOD SOURCE FOR INDONESIANS. THAT'S ANOTHER REASON IT'S DISAPPEARING.

GLOSSARY

adaptation: A change in a living thing that helps it live better in its habitat.

carrion: A dead, rotting animal.

chemical: Matter that can be mixed with other matter to cause changes.

equator: An imaginary line around Earth that is the same distance from the North and South Poles.

fertilization: The act of adding male cells to female cells to produce new plants.

habitat: The natural place where an animal or plant lives.

humidity: Moisture, or wetness, in the air.

insect: A small, often winged, animal with six legs and three main body parts.

photosynthesis: The process by which a plant turns water and carbon dioxide into food when the plant is exposed to sunlight.

release: To let something go.

reproduce: When an animal or plant creates another creature just like itself.

tuber: A short, thick, round stem that is a part of certain plants, that grows underground, and that can produce a new plant.

FOR MORE INFORMATION

BOOKS

Braun, Eric. *Corpse Flower vs. Venus Flytrap.* Mankato, MN: Black Rabbit Books, 2018.

Einstein, Tamara. *Weird Plants.* Vancouver, Canada: KidsWorld, 2021.

Levy, Janey. *Freaky Stories About Plants.* New York, NY: Gareth Stevens Publishing, 2017.

WEBSITES

Darth Vapor: The Corpse Flower
californiasciencecenter.org/exhibits/life-beginnings/corpse-flower
Check out a video about the pollination of this cool plant.

The Titan Arum
www.chicagobotanic.org/titan/faq
Find more answers about the corpse flower on the Chicago Botanic Garden site.

INDEX